# Tsunami!

## Death Wave

by Margo Sorenson

**Perfection Learning® CA**

Cover Photo:   National Oceanic and Atmospheric Administration
Illustrations:   Larry Nolte
                 Kay Ewald

# Dedication

For Jim, Jane, and Jill,
who were always ready to hear more tsunami facts.
For Ben, Joe, and Jak,
for all their great help and enthusiasm for tsunamis.
For Mike Blackford and Chip McCreery, for their time,
their patience, and their willingness to answer many, many questions.

# About the Author

Margo Sorenson was born June 17, 1946, in Washington, D.C. She finished her school years in California, graduating from the University of California at Los Angeles. She taught high school and middle school in Hawaii for ten years, while raising a family of two daughters. Mrs. Sorenson is now a full-time writer of young adult literature.

Living in Hawaii gave Mrs. Sorenson firsthand experience with the threat of tsunamis. Her house was on the beach, so when earthquakes hit across the Pacific and tsunami warnings were issued, her family had to evacuate. Mrs. Sorenson enjoyed researching and writing this book because she learned so much about tsunamis. She confesses that if she had written this book before she moved to Hawaii, she never would have lived where she did!

Mrs. Sorenson now lives in Minnesota with her husband but travels back to Hawaii almost every year. When she isn't writing books, Mrs. Sorenson enjoys sports, watching the weather channel, and writing letters to her friends in Hawaii.

Printed in the United States of America. For information, contact
Perfection Learning® Corporation, 1000 North Second Avenue,
P.O. Box 500, Logan, Iowa 51546-1099
ISBN  0-7891-1953-6 Paperback
ISBN  07-807-6146-4 Cover Craft®
8 9 10 11 12 13 PP 09 08 07 06 05 04

# Table of Contents

This tsunami at Laie Point on the Island of Oahu, Hawaii, was generated      NOAA
by an earthquake in the Aleutian Islands on March 9, 1957.

# CHAPTER 1

# Here It Comes!

Two children played in the sand on the sunny beach. The little boy dug with his blue shovel, while his sister splashed at the ocean's edge. She scooped up water in her red plastic bucket.

The children didn't know death was on its way!

At the harbor, people were busy. Men drove trucks to the docks. Dockworkers loaded ships. Ship captains checked off lists. The ships readied to leave.

The workers didn't know death was on its way!

The town lay next to the harbor. A tailor stood in the doorway of his shop. A butcher wrapped some meat for a customer. Lawyers sat in their offices doing research. Taxi drivers honked at street corners. People strolled the streets. A woman walked her dog.

Death was coming for them too.

In a house nearby, a father cleaned the kitchen. A boy trimmed a hedge. A grandmother read a book. The mail carrier delivered letters.

Death was coming for them too.

Far, far away—thousands of miles away—the ocean floor

moved. A huge plate of the earth's crust shifted in the wet darkness. It was an earthquake.

The undersea world jolted. The ocean floor heaved under thousands of feet of dark, dark water.

The massive pressure of the moving earth forced the water upward. Shock waves rolled up through the dark water. Odd sea creatures scattered.

Thousands of tons of water rolled above the quake. It surged up toward the light. Giant waves formed.

The giant waves headed for the beach, the harbor, and the town. They glided through the open ocean.

The waves rolled, one after another. Huge columns of water rolled hundreds of miles apart—silent and deadly.

In the open ocean, the death waves looked like ordinary waves—hardly higher than the other waves. Who would know they were moving at jet speed?

An ocean liner sailed through the waves. The captain and crew didn't notice any difference.

Laughing passengers played games on the ship's deck. Some looked over the rail. The deadly waves passed the ship, one by one. But the passengers noticed nothing unusual. The ocean hid its secret well.

On and on the waves powered through the water at incredible speed. At 500 mph, the waves sped toward the shore. They were bringing death to the land.

An hour had passed since the undersea quake. The water at the ocean bottom still swirled. The sea life still scattered and fled. But the death waves were well on their way.

On land, people had felt a slight tremor. It was from the far-away earthquake. So no one thought much about it.

All was safe, after all. The houses were safe. The shops and offices were untouched.

In the harbor, the loading went on as before. Fishing boats traveled outside the harbor. An oil tanker sailed toward the town. And the children built their sand castle on the beach.

All the while, the huge, secret waves rolled closer to the land. As the ocean became shallower, there was no room for the huge waves. Their energy compressed, forcing them upward. The waves grew higher and higher.

The first wave became 4 feet, then 6. It grew to 10, then 20 feet high. The wave's speed slowed—from 500 mph to 40 mph.

Other waves piled up behind the first wave, which was now 10 miles from shore. It rolled toward the land like a giant wheel of water.

The rolling motion of the wave pulled at the ocean in front of it. This caused the water to drain away from the shore toward the giant waves.

On the beach, the children stopped their play. "Look at the water!" the little girl cried. She laughed. "It's going bye-bye." She waved at the water.

The little boy looked up from his sand castle. He put down his shovel and stared at the ocean. It was disappearing.

The water made a strange sucking noise. It hissed loudly. Some strange force pulled it back away from the land. A few fish wriggled on the exposed sea bed. Shells gleamed on the wet sand.

Then he smiled. "It's a bathtub! Somebody pulled the plug," he joked. They giggled together and went back to their sand castle.

In the harbor, the water level suddenly began to drop. The dockworkers stopped their work.

Tied-up fishing boats snapped their lines. They rode backward toward the ocean, pulled by the undertow out of the harbor.

"What's happening?" they asked each other. Their eyes were wide with fright.

"No, it couldn't be," said one. He put down the bales he was loading and looked at his partner. "Could it be?" he asked.

Everyone looked toward the ocean. In the distance, a blue-green swell massed on the horizon. It rose 20 feet. It surged toward them at 40 mph.

A lawyer looked out her window at the harbor. She froze. The harbor was nearly empty of water. Stones on the harbor bottom gleamed wetly.

She peered out to sea. In the distance, a giant wave loomed. Sunlight sparkled on its huge crest.

"No!" she screamed. Was she up high enough? Was her building far enough away from the harbor? How far could the wave reach?

What should she do? Could she run for higher ground? Would she be safe?

Within 20 minutes of the harbor emptying, the first wave crashed. An oil tanker ¼ of a mile out felt the wave's power.

The captain couldn't steer the tanker, and it surged into the pier. The pier gouged a giant hole in the tanker. Oil poured out.

Tons of water smashed onto the ships and boats in the harbor. Many splintered like matchsticks. Oil spilled from the tanker, and boats caught fire.

The giant wave picked up gallons of burning oil. The flaming current roared inland to the town.

The first wave crushed the pier. The docks collapsed like toy building blocks. The wave rolled onward. It carried the skeletons of boats. Burning debris floated with it.

The first wave surged into the town. It battered against buildings. They collapsed into the roiling foam. Buildings caught fire.

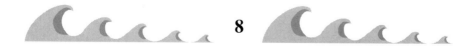

Screams filled the air. The dockworkers vanished, sucked under by the wave.

People drowned by the dozens. Those who had seen the first wave were already gone. Thousands of gallons of water had crushed them.

At the beach, nothing was left. The sand castle was only a memory. A sand shovel and a red bucket floated hundreds of yards inland—all by themselves.

Then the sea began to withdraw. It pulled back toward the ocean again. Logs, pieces of buildings, railroad ties, everything began to surge back out to sea.

In the town, a few dazed survivors wandered around. "I can't believe it," a woman said. She held her forehead in shock. Another woman sobbed. From the soaked rubble around them, they could hear moans.

A man ran down a street. He rushed toward the harbor. He held a camera. "I've got to get a picture of this," he said. He hurried to aim his camera at the devastation.

In the distance, a second wave rose up. Its dark green wall rolled menacingly toward the town.

The second wave crashed, harder than the first. And then all was silent. The man with the camera would take no more pictures. His camera floated by in the sea surge.

The next wave smashed onto the shore. It rolled inland, carrying burning oil with it. More buildings caught fire. A few survivors clung to floating debris. They screamed as the burning oil engulfed them.

The swift current tossed lumber around. Jagged pieces of iron churned in the water. They battered against buildings. More buildings collapsed. Steel twisted. Slabs of concrete were ripped from their foundations.

Moans and screams filled the air. Distant sirens began to

whine. Emergency vehicles roared toward the town. Fire engines honked along the inland highways. Police cars flashed their lights. Ambulances raced toward the devastation.

The next wave towered over the harbor. The force of the waves behind it piled it even higher. Its deep blue face leered at the town. With a roar, the wave surged through the harbor.

The wave powered into the town at 40 mph. It raged with destruction. With a murderous sweep, it cleared all in its path.

Police cars halted. Their sirens screamed in the silence. Lights flashed. Fire trucks stopped. Ambulances braked. Emergency and rescue crews huddled at the edge of town.

"How bad is it?" one paramedic asked. He looked toward the ruined town.

The officer frowned. "It's terrible!" he said. "The worst I've ever seen!"

Little groups of rescuers bunched together. They stared at the destruction. They talked in whispers.

A voice on a loudspeaker warned people not to enter the town. There were probably more waves coming. But no one knew how many.

Finally, the all-clear sounded. Rescue vehicles and personnel began their sad tasks.

Bizarre sights greeted the rescuers. Boats were beached on top of smashed buildings. Uprooted trees lay inside house skeletons. Fires burned everywhere.

The heavy smell of oil and smoke stung the rescuers' nostrils. Their eyes watered from the smoky air. They could hear moaning.

"Help! Over here!" a voice yelled. A man clung to a telephone pole. By a miracle, it still stood. People rushed to rescue him.

Injured people were taken to hospitals in ambulances. But there were few survivors. Grim-faced men carried bodies from the scene of destruction. The long process of identifying bodies began.

Rescuers combed the wreckage. They hoped to find survivors. But over days, the death toll grew . . . fifty . . . one hundred . . . two hundred were lost. Hundreds more lay injured in hospitals. Hundreds were homeless.

Cleanup crews began clearing wreckage. Cranes rolled over muddy ground. Bulldozers cleared foundations. Men and women drove tractors and trucks, hauling debris.

One worker stopped. Something had caught his eye. He shut down the engine and stepped down from the cab.

Then he reached down to the mud. The worker held up what he had found. Quick tears stung his eyes.

It was a red, plastic bucket.

A sandy beach is all that remains. Waves removed all traces of Riangkroko, Indonesia.

NOAA

## CHAPTER 2

# From Inside the Secret Earth

### What is a tsunami?

A **tsunami** causes death and destruction. Its power terrifies those who see it. Just the word triggers fear in people in coastal areas.

What is this force of nature that has killed over 50,000 people since 1888 and caused millions of dollars in damage?

*Tsunami* is Japanese for "harbor wave." A tsunami (SOO-<u>NAH</u>-MEE) is a huge wave. Some are more than 80 feet high.

Tsunamis actually come in a series. They never come one at a time. As many as 7 to 12 tsunami waves may hit a coast.

Many think tsunamis are regular waves—just bigger. But that isn't true. Others mistakenly think tides make tsunamis. That is wrong too. **Tidal waves** are different. They happen naturally due to **gravity.**

The positions of the earth, moon, and sun cause gravity. So tides can be predicted, but tsunamis cannot. Tides are regular and have rhythm. Tsunamis have neither.

Even the world's highest tide, 40 feet in the Bay of Fundy, Canada, isn't a tsunami. Everyone knows it's coming. It's a natural event. It comes little by little, over a 12-hour period. It doesn't happen all at once.

Some people think a storm can cause a tsunami. That's also not true.

Huge storm waves are called *storm surge,* and they can be predicted. Weather experts can track a storm. They can tell where a storm surge is likely to happen.

People have time to prepare. Owners haul boats from the water. People often sandbag their property. If necessary, they can evacuate.

But no one knows when tsunamis will occur. No one can predict them.

Imagine a series of huge waves. And no one knows when they might come. Often, there is no warning.

## What causes tsunamis?

Tsunamis aren't caused by tidal changes. They're not made by storms either. Forces deep in the earth cause tsunamis.

*Geology* is the study of the earth. And geologic events cause tsunamis. Earthquakes, volcanic eruptions, and landslides are examples of geologic events. But since we can't see deep inside the earth, no one knows exactly what's happening.

A **landslide** can cause a tsunami. During a landslide, tons of mud, rock, or ice slide down a mountain into a lake or a reservoir. This force can trigger a tsunami.

The waves spread through the body of water. They damage the shoreline and roll over the tops of dams. Flooding threatens areas downstream too.

Sometimes, a landslide happens under the sea. There are underwater slopes leading to the ocean's bottom. On these slopes, huge deposits of soil or sand can suddenly break free. Gravity finally has its way. The tons of earth begin sliding downward.

An erupting **volcano** can also start a tsunami. If a volcano erupts on the ocean floor, the ocean bottom explodes. The force

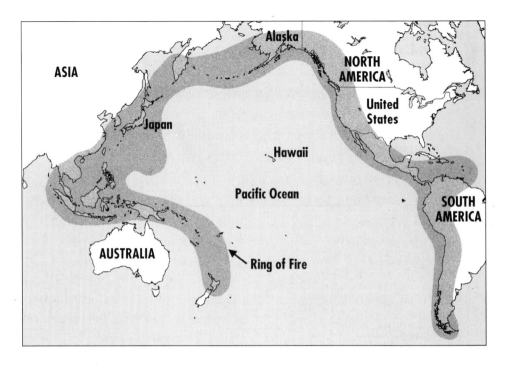

of the explosion launches a tsunami.

Many volcanoes, though, are not underwater. They are on the land. But those near water can cause tsunamis too.

As a volcano erupts, some of the eruption cools in the air. It can become volcanic ash and rocks.

The ash and rocks fall back down to earth. Gravity pulls them down the sides of the volcano. The ash and rocks slide into the water. This can cause a tsunami as the volcanic material pushes the water aside.

The countries circling the Pacific Ocean have more volcanoes than anywhere in the world. This area is called the ***Ring of Fire.*** Of the 2,500 volcanic eruptions recorded, 2,200 of them have been in the Ring of Fire.

The Ring of Fire follows the western edges of North and South America. Then it loops under the ocean. It follows the coasts of Australia and Japan and runs up to Russia.

Most tsunamis, though, start with **earthquakes.** And most earthquakes that trigger tsunamis occur around the **Pacific Rim.** These earthquakes occur in areas called *subduction zones.*

There are many subduction zones in the Pacific Rim countries. They all lie near coastal areas.

What is a subduction zone? Picture an M & M candy peanut. It has three layers—the peanut, the chocolate, and the shell. Now pretend it's the earth.

The earth is made up of three layers—the **core** (inner and outer), the **mantle,** and the **crust.** The core (peanut) is made of molten metals and is about 2,150 miles deep.

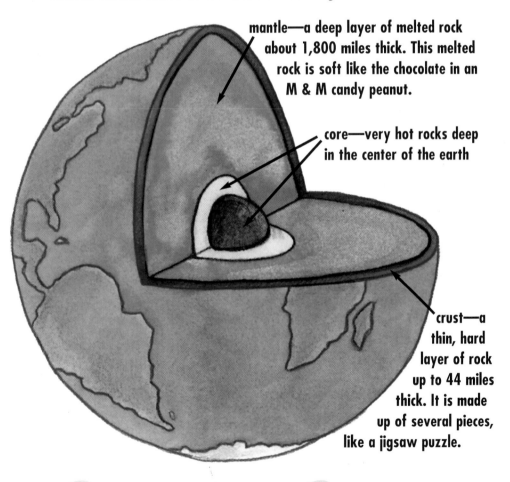

mantle—a deep layer of melted rock about 1,800 miles thick. This melted rock is soft like the chocolate in an M & M candy peanut.

core—very hot rocks deep in the center of the earth

crust—a thin, hard layer of rock up to 44 miles thick. It is made up of several pieces, like a jigsaw puzzle.

The mantle (the chocolate) lies on top of that. It's about 1,800 miles deep.

Scientists think the mantle flows, but they want to learn more. So they are conducting **sonar experiments.**

The crust (candy shell) is the topmost layer. It lies in sections, or **plates,** on the mantle. It averages 20 miles deep. But in some places it lies 40 miles deep. The plates float on the mantle like lily pads on a pond.

The crust or plates of the ocean floor are very heavy. A subduction zone is created when one plate overrides or subducts another plate. The lower plate is pushed downward into the mantle, where it melts.

Why is this dangerous? The lower ocean plate sinks a few inches a year. When it sinks, it pulls and scrapes against the continental plate above it. This creates stress.

The stress builds over hundreds of years. Finally, the plates jerk apart, causing an earthquake. An undersea quake is "like a huge number of atomic bombs going off," says one **oceanographer.** The earthquake can raise or lower the ocean bottom.

Generally, only quakes under the sea start tsunamis. In other quake faults on dry land, crustal plates slide past each other, not up and down. And it is the up-and-down movement that creates a tsunami.

## How does the tsunami form?

Imagine dropping a rock into a pool. Ripples spread out in rings from where the rock is dropped. In a way, each ripple is like a tsunami. But there are dangerous differences!

When the ocean bottom rises or sinks, a huge column of water begins to move up or down. The movement occurs from the sea floor to the surface of the ocean.

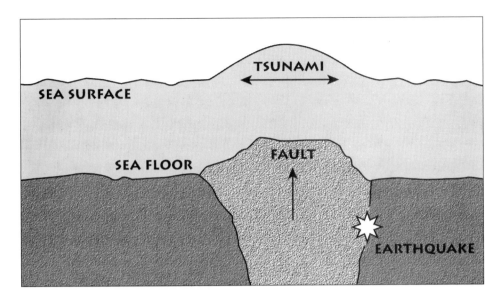

If the sea floor rises, the surface water will rise. If the floor lowers, the huge column of water will drop.

This is why tsunami waves are much more powerful than normal waves. The regular wave's power is only a few feet down into the ocean. A tsunami's current extends all the way to the ocean floor.

Think back to the pool again. When the rock is dropped, the water ripples move toward the side of the pool. The ripples hit the side, splash, and then roll back to hit the other side. The ripples slosh from side to side.

The same thing happens in a tsunami. The rising and falling of the water sets the tsunami in motion. The ocean keeps rising and sinking in huge waves all the way to the nearest shore. Tsunamis can rocket back and forth for days between continents.

## How does the tsunami travel?

Each tsunami wave swells up and down on its way to land. In the open ocean, it can travel 500 mph.

It's hard to tell the speed and height of a wave in the middle of the ocean. That's because other waves swell across the ocean too. So tsunamis are hard to recognize. But once the tsunami wave rolls closer to shore, we can see it.

Also, as much as 600 miles might separate one tsunami crest from the next. And the slope of the tsunami crest is gradual. It might get higher only by an inch or so each mile. This also makes a tsunami wave hard to recognize.

Tsunamis travel far. A large earthquake can send a tsunami traveling across a whole ocean. It rolls thousands of miles. It keeps its power.

An earthquake in Chile can send a tsunami to Hawaii in ten hours. It will reach Japan in 24 hours. Normal waves lose power after a few miles.

## Tsunami Travel Times to/from Honolulu

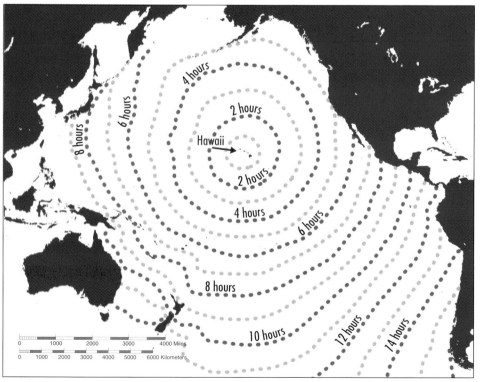

A view of the complete destruction of Pagaraman, on Babi Island. About NOAA 700 people were killed and over 100 reported missing.

The ocean floor can change the tsunami's direction. **Submarine canyons** and mountains force the tsunami on a different course. The wave's power changes direction.

During the 1992 Indonesian earthquake, villages were caught off guard. The tsunami began north of Babi Island. It should have hit the north shore villages. But the tsunami pounded the southern villages. Scientists think changes in the sea floor made the tsunami swerve around the island.

## What happens when a tsunami gets closer to shore?

Closer to shore, the ocean gets shallower. The bottom of the ocean gets closer to the surface. This squeezes the energy of the tsunami. It can't go down. So it goes up instead.

Think of squeezing a blown-up balloon. The more you squeeze it at one end, the bigger it gets at the other. This is one reason tsunamis rise so high close to shore. The bottom of the ocean is "squeezing" the wave up into the air.

Near the shore, the power of the tsunami fights against the ocean floor. This friction against the bottom causes the tsunami to slow down. From 500 mph, it might slow to half that, then half again, in just a mile. A tsunami hits the shore at about 40 mph.

You would think that slowing down would make the tsunami less dangerous. But it doesn't work that way.

Think of a line of cars following closely behind each other. What happens if the first car stops suddenly? All of the other cars crash into each other and pile up.

It's the same with the tsunami waves. The water behind the slower waves in front is still traveling at a higher rate of speed. So the fast water behind piles up against the slower water in front.

The pressure from behind causes the waves to grow in height. By the time each wave hits shore, it may rise up to 30 feet high.

Imagine a wall of water 30 feet high slamming into shore. For every 5 feet of coastline, a large tsunami can dump more than 100,000 tons of water.

But that's not all. Think of the wave as a wheel. The wheel turns around and around. It rolls closer to the land. As it turns, the bottom of the "wheel" pulls at the water next to the land.

Beachgoers know this feeling. Just before each new ocean wave breaks, the water rushes back out to sea.

With a tsunami, the pull is much stronger. A tsunami sucks all the water away from shore. Tsunamis have drained whole harbors.

Then the sea rolls back to meet the next wave. Debris, lumber, and boats are sucked back out to sea.

The destruction from tsunamis is widespread. The first wave can sweep away buildings, bridges, docks, piers, and trees. It can even carry ships for miles inland.

Then the next waves power into shore. These waves use what has been destroyed as battering rams. Huge pieces of lumber smash into buildings still standing.

Oil from ships catches fire from engine sparks. The burning oil rushes inland. Buildings catch fire. Trees uproot. The strong current rages. It shoves trees into more buildings. The torrent rushes on. It sweeps everything in its path. Many people die.

The tsunami has done its awful work.

April 1, 1946. Tsunami is breaking over the pier at Hilo, Hawaii. Man in foreground became one of 173 deaths. Photo was taken from the *Brigham Victory,* which was in the harbor.

NOAA

# CHAPTER 3

# Killing the Demon Wave

## How can geologists help?

We can't prevent tsunamis. But we can do things to protect lives and property.

All over the world, scientists are studying tsunamis. **Geologists** act as "earth detectives." They learn about soils and where it's safe to build.

Weak soil breaks apart easily and is dangerous near a coast. A tsunami could wash away weak soil under a building. So geologists advise against building in certain areas.

Geologists also study the history of a site. They find clues about what has happened in past years. Has water flowed over the land? Were there tsunamis? Were there storm surges?

The geologists look for signs that water has at one time covered the land. Such signs incude sand particles far from the ocean or little bits of ancient sea plants in the soil. Even tiny marine fossils can be clues.

When scientists find these signs, they can tell that a tsunami once surged ashore. So it's probably not a safe place to build.

For example, geologists discovered interesting clues in the sand and sea grass in Puget Sound, Washington. The fossils and sea grass told them that a tidal wave had swept into the area

1,000 years before. That was about the time a huge tsunami hit Japan. Scientists believe that Puget Sound was pounded by the same tsunami. That means the tsunami rolled from Japan to North America.

American Indians in the Puget Sound area tell legends about flooding from the sea. It sounds like they're talking about tsunamis. There has been no tsunami since, though.

Today, the city of Seattle is built on Puget Sound. So far, it has been safe.

Some scientists, though, worry about the nearby Cascadia subduction zone, which lies under the ocean. It borders the northwest coast of the U.S. and stretches from Vancouver to northern California. It could cause a huge earthquake and tsunami. And there is no way to predict when or if it might happen.

People who need to build in the ocean itself should also talk to geologists. Strong waves can hit offshore drilling platforms, docks, and piers. They can wash away man-made breakwaters and seawalls.

Geologists also can design objects that will help protect the shore from strong waves. Builders should know how shorelines react to waves. This knowledge could help seawalls stand up against smaller waves.

## How can oceanographers help?

Oceanographers also help in the study of tsunamis. Oceanographers study ocean currents, movements, and other data. They set up tide sensors and gauges all through the Pacific Ocean and the Pacific Rim. These tide sensors send data to the Pacific Tsunami Warning Center and the Alaska Tsunami Warning Center.

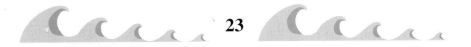

## How can seismologists help?

Remember, earthquakes under the sea cause tsunamis. And **seismologists** study earthquakes.

Seismologists study the movement of the earth's plates. They map out faults on the sea floor.

Knowing more about earthquakes can help people prepare for tsunamis. People living on nearby coasts can make emergency plans. Their civil defense agencies can plan for dealing with tsunamis. Their governments can be ready.

November 4, 1952. Aerial view of the north shore of Oahu, Hawaii.　NOAA
Photo shows the fourth wave climbing toward the beach houses.

## What are tsunami warning systems?

Scientists work hard studying tsunamis. They want to save lives. Many countries and agencies share information they learn about tsunamis.

In 1946, a tsunami warning system was set up. The Intergovernmental Oceanographic Commission (I.O.C.) set this up. It is called the **Operational Tsunami Warning System.** Now over 25 countries belong to the system. The countries all share information.

The Pacific Tsunami Warning Center (P.T.W.C.) in Honolulu is the system's headquarters. There is also a Tsunami Warning Center in Alaska.

The P.T.W.C. is a huge command center for tsunami control. The P.T.W.C. gets information from many sources. It's on-line with many agencies around the world. The staff also remains in touch with news reports.

The P.T.W.C. receives data about earthquakes from the world's stations. Any earthquake above 6.5 on the **Richter scale** triggers an alarm. Within five minutes, the P.T.W.C. alerts all affected countries.

The P.T.W.C. also gets data from tidal gauges or sensors all over the Pacific. Tidal gauges float in the ocean. They tell how the water level is changing. If water is rising quickly, the P.T.W.C. notifies others around the Pacific Rim.

The P.T.W.C. has a special computer program that computes tsunami travel time. It uses earthquake data and tidal sensor data.

The program also uses information about ocean depth and what the ocean bottom is like. Underwater canyons and landscape can affect where a tsunami will hit. They can channel the tsunami waves. So tsunamis may not travel in a straight line.

Tsunami destruction in Alaska                                    NOAA

Using necessary data, the computer program predicts how fast a tsunami will travel. How many hours until it hits Hawaii, then Chile, Alaska, and Indonesia? A tsunami can travel the Pacific in less than 24 hours.

The computer program also predicts where a tsunami might hit. In this way, the scientists at P.T.W.C. help to save lives. They warn potential victims.

Another group of scientists form the International Tsunami Information Center (I.T.I.C.). It shares information about tsunamis, educates people about the dangers, and helps train scientists.

The I.T.I.C. also monitors the tsunami warning system. Many countries meet with the I.T.I.C. every two years. They decide how to keep the world informed about tsunami dangers.

# What is a tsunami earthquake?

Scientists still have questions about tsunamis. It's not hard to predict a tsunami when an earthquake of 6.5 or more hits. But some smaller earthquakes cause terrible tsunamis. And scientists wonder how tsunamis from small earthquakes can cause so much destruction.

For example, in Nicaragua in 1992, many people on the coast hardly felt the earthquake. Then 45 minutes after the quake, danger rushed ashore. A tsunami almost 33 feet high raged onto the coast. The wave drowned 116 people.

Hiroo Kanamori, a scientist at Caltech, has studied this problem. He calls this a **tsunami earthquake.**

Kanamori has studied tsunami earthquakes for more than 20 years. When he began studying them, instruments measuring earthquakes weren't very advanced. They only measured short-period waves of action in the earth.

Short-period waves jerk the surface hard. Everyone notices these jolting earthquakes.

But what about the earthquakes no one felt? Kanamori guessed long-period waves caused these. But no one could measure them.

Now instruments can measure long-period waves of action too. For example, the Nicaraguan earthquake measured 7.0 with old instruments. Just the short-period waves were measured. With new instruments, the earthquake came up as 7.6.

This means that the Nicaragua quake was really almost ten times as strong as people first thought. No wonder it caused a tsunami!

**Slow motion slip** is what Kanamori calls this kind of earthquake movement. He thinks this kind of quake makes great energy. But it hides its real power.

Mr. Kanamori suggests that one oceanic plate slides slowly under an oceanic plate next to it. But ocean sediments soften the slide's impact. These cause long-period waves. The plates move just as far. But they don't jerk as much.

Imagine sliding one dinner plate over another. You would notice the loud noise. You might even break the plates.

But imagine a layer of jello in between the plates. You wouldn't hear anything.

This is what happens in a slow-slip earthquake. The plates still move. An earthquake happens. But the earth doesn't jolt as fiercely.

Kanamori says scientists must analyze long-period waves and alert people about the slow-slip quakes. Otherwise, no one will notice the earthquake happening. And at sea, a deadly tsunami could be forming.

In 1992, a station in Tahiti picked up the long-period waves from the Nicaraguan earthquake. Tahiti is 4,000 miles away. Seismologist Emile Okal saw right away that it could be a disaster for Nicaragua.

But it was already too late. By the time Tahiti got the signal, the tsunami had destroyed the Nicaraguan coast.

It took 25 minutes for the signal from the quake to travel to Tahiti. Unfortunately, the tsunami had to cover only 40 miles before hitting the Nicaraguan coast.

## How are people warned?

The hard work of all these scientists is important. People must know about tsunamis. Each country in the Pacific Rim must take care of its own population.

The P.T.W.C. and other agencies warn the areas and countries in danger. **Civil defense** and national agencies spring into action. Each area has its own plans. News programs on radio and TV blare out the news.

Some places, like Hawaii, alert civil defense. There, sirens blare out for a tsunami warning. Everyone must evacuate.

People in Hawaii know if they live in a danger zone. Tsunami danger zones are even published in telephone books. People living in these areas must buy extra insurance too. When you buy a house in an area like this, you have to sign papers that you know you're at risk for tsunamis.

## What can you do about tsunamis?

If you live by the ocean, you should know about tsunamis. You should have an evacuation plan. You should be ready.

Also, if you visit the ocean, you should be aware of tsunamis. Find out how people are alerted for tsunamis where you're visiting. Is it by a siren? Is it by police helicopters?

If you feel an earthquake on the beach, don't wait for a warning. Run to higher ground. Run if you see the waves being pulled out quickly. A tsunami may be on its way.

When a tsunami warning sounds, get to higher ground quickly. Listen to the news if you can. News people can tell you where you should go. Listen carefully. Don't panic.

Don't think you can watch the tsunami. Remember, *if you can see it, you can't outrun it.*

Wait until authorities sound an all-clear before you return. The ocean may look safe. But tsunamis come in series. As many as 24 hours can pass before all the waves have died down.

The danger is too great. Don't take a chance!

# CHAPTER

# Tsunami-Tamers Talk About Tsunamis

Mr. Mike Blackford sits up in bed. It's four a.m. and he's wide awake. He's on duty, and his pager is beeping.

Quickly, he runs the short distance to the main building. He flings open the door.

The high-pitched whine of the alarm greets him. Then it breaks into a loud "Beep-bop-beep-bop-beep-bop." It sounds like an emergency vehicle. But it is the alarm for the Pacific Tsunami Warning Center.

## How does the Pacific Tsunami Warning Center work?

Mr. Blackford and four other scientists staff the P.T.W.C. around the clock. Three of them live at the center in houses. The other two live elsewhere. They share a fourth house at the center when they are on duty.

The entire staff works two-day shifts, 24 hours a day. They carry pagers all the time. The pagers are tuned into the warning system.

The alarm means an earthquake has hit. Somewhere in the Pacific Basin, the earth has jolted. A deadly tsunami might form.

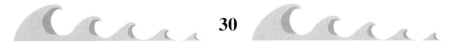

Mr. Blackford checks the computer. A display of **seismic** data glows from the screen. Which seismic stations have gotten data? Where is the quake?

Mr. Blackford checks the stations showing signals. Then he logs into **N.E.I.C. (National Earthquake Information Center)**. Now he can get data from hundreds of stations.

Quickly, he logs the data from N.E.I.C. Then he sends all the data to the P.T.W.C. computer. Now he knows the source of the quake.

It shows as 7.5 on the Richter scale. It's a big one. Right away, Mr. Blackford uses the P.T.W.C. system to generate tsunami watches and warnings. The P.T.W.C. system sends these messages all over the Pacific.

Only a few minutes have passed since the quake hit. But time is critical.

The places in danger get the message. Tsunami warning! Tsunami watch!

Around the Pacific, people spring into action. Countries get to work. Cities and towns are alerted. They contact civil defense. Police are called. Sirens sound. Coastal areas are evacuated.

Where will the tsunami hit? How hard will it hit? When will it hit?

Mr. Blackford sits in front of a computer screen. He feeds data into a special tsunami program. The program will answer questions about the wave.

At the source of the quake, the water level changes fast. It may rise as high as ten meters in minutes.

So Mr. Blackford checks water level data from over 100 sensors around the Pacific. Satellites relay the information to him. They show whether the ocean level is changing.

Information speeds to the center from many stations. News reports, phone calls, and on-line information stream in. He enters more data about the quake.

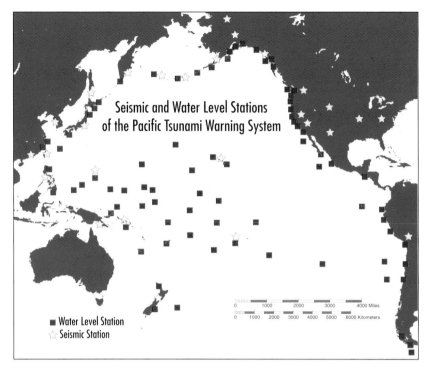

Seismic and Water Level Stations of the Pacific Tsunami Warning System

■ Water Level Station
☆ Seismic Station

The special tsunami computer program spits back numbers. The program knows the ocean's depth. The program also knows the coastline shapes.

The program uses these and other data Mr. Blackford keeps adding. It tells where the tsunami will travel first. And it figures how soon the tsunami will get there.

All areas within three hours of the first wave are on tsunami warning. Areas within three to six hours are on watch. This isn't much time.

Mr. Blackford has decisions to make. When will he lift the warning? Will he upgrade the watch to a warning? Will other places need to be on warning too?

Every hour, Mr. Blackford checks data. How high is water rising? How is it changing? How many aftershocks are there?

Mr. Blackford explains, "I watch to see if water is rising 1½ to 2 meters at 300 to 500 miles from the source area. If so, I continue a warning.

"If it's rising over 2 meters, I send a Pacific-wide warning." He adds, "If there are damage reports, a warning is continued."

Mr. Blackford knows the P.T.W.C. saves lives. "I like it when we can issue warnings smoothly. It's important everyone gets the message."

## What if the P.T.W.C. breaks down?

This is life-and-death work, so Mr. Blackford has concerns too. The breakdown of the P.T.W.C. system is his worst nightmare. Lives could be lost. Millions of dollars of damage could be done.

In the event of a breakdown, Mr. Blackford would call the other tsunami warning center in Alaska. It covers Alaska, British Columbia, Washington, Oregon, and California. Each center backs up the other when there are problems.

Mr. Blackford would call other agencies and centers for information and help. He would have to gather and compute data himself. It would take more time. And time is important.

Luckily, this is not likely to happen. Many backup systems are in place.

## Why are tsunamis so powerful?

Mr. Blackford says many people don't realize a tsunami's strength. For example, a six-foot tsunami might not sound bad. After all, the tide in San Francisco Bay rises and falls between six and eight feet.

But there's more to it. The tide rises and falls that much over a 12-hour period. It comes and goes quietly. The change is slow.

But a tsunami changes ocean levels fast. It could rush in at 6 feet every 15 to 30 minutes. The power of the rushing waves is very strong.

The pulling power of the waves is strong too. Remember, tsunamis also pull the ocean away from shore. And the strong current going out also causes great damage. Then the tsunamis barrel into the shore again.

And tsunamis are more than just one wave, Mr. Blackford says. Sometimes the first one isn't the worst. People forget that.

Mr. Blackford has advice. "If you hear a warning that a tsunami is coming, believe it," he says. "Just because you can't see it doesn't mean anything. Get to high ground as soon as you can."

He also warns about being on a beach and feeling an earthquake. "If it's strong enough to make you fall, run. Don't even look back. Run for higher ground."

Tsunami waves in the open ocean are only a few feet high. But the wave height increases as the tsunami nears the coastline. The wave energy is pressed into a shorter distance in the shallow water. This creates tall, life-threatening waves that batter the coastline.

Ocean

Ocean Floor

## How can tsunami prediction be improved?

Scientists from all over the world sit at a huge table. Twenty-six countries are represented. It's been two years since they last met. They are discussing tsunamis. The warning systems and public education are being evaluated.

What can we do better? Are we warning people in time? What tsunamis have happened in the last two years? What did

people do? How did the systems work? What new systems can we use?

Sitting at the table is Dr. Charles McCreery. He is the director of I.T.I.C., the International Tsunami Information Center. It is located in Honolulu, Hawaii.

Dr. McCreery listens and shares ideas with other countries. Educating people about tsunamis is his job. He writes down new tasks for the I.T.I.C.

Dr. McCreery wants to save lives. He says public education is one of the best ways to do that.

Most of the deaths from tsunamis happen right away. If a tsunami comes from an earthquake nearby, there may not be time to get out a warning. Even if the earthquake happens thousands of miles away, warnings might not reach remote areas. People must learn the signs of danger for themselves.

If people are near the ocean, they run a risk, Dr. McCreery says. They could be overrun by tsunami waves without warning.

Because tsunamis can travel at 500 mph in the ocean, their speed is a danger. An earthquake may have happened hundreds of miles away. But the tsunami could arrive at a coast without warning.

Dr. McCreery would like people living along the coasts to be better educated. He would also like to see more and better tsunami sensors—ones that could relay their data right away. Warning centers need the data right as it happens.

Dr. McCreery also wants people to enjoy the beach and the ocean. But they must act safely. If they feel a strong earthquake and are near the shore, they need to get to higher ground. If the water rushes out of a bay or inlet, they need to run to higher ground.

Scientists also need better data to predict tsunami heights. One proposal is to use deep ocean-pressure sensors to measure

April 1, 1946. People flee as a tsunami attacks downtown Hilo, Hawaii. NOAA

tsunami waves. These sensors have been successfully tested. But there was no good way to relay the data. It had to be picked up by a ship. This was too late for the warning system.

A new design would relay the data back to the warning centers through a satellite. These new sensors would be placed in critical locations around the Pacific. This could help save lives.

Dr. McCreery tells a tragic story about a tsunami. In 1946, a huge tsunami hit the island of Hawaii. It came from an earthquake in Alaska, thousands of miles away.

The tsunami raged through Hilo. But it also hit other coastal towns.

At Laupahoehoe, an elementary school sat on a spit of land. It was 7 a.m. when the tsunami roared over the land. Some of the children were just arriving at school. The school was destroyed in minutes.

Many children were washed out to sea. Sixteen students and five teachers were killed. Some who hung onto debris were later saved. It was a tragic day. If they had been warned, would they be alive now?

Today, a plaque marks the spot. Tourists often stop by the road and gaze at the site.

Without systems like the P.T.W.C. and the I.T.I.C., tsunamis would bring even more destruction. Geophysicists like Mr. Mike Blackford and Dr. Charles McCreery are important. Their jobs are exciting and very critical.

April 1, 1946. Hilo, Hawaii. Huge waves snapped wood-frame houses and stores and left debris everywhere.

Hawaii State Archives

# CHAPTER 5

# Tales of Terror

## What do legends tell us about tsunamis?

It is important to find out about past tsunamis. Studying what has happened in the past can help people become better prepared today.

Scientists may need to search for clues to learn about the history of tsunamis. In chapter 3, you read how scientists study the earth. They find fossils, sand, and other signs.

Ancient legends can also provide clues. For example, an American Indian legend helped scientists theorize about the Puget Sound tsunami 1,000 years ago. Learn more about this legend on page 46.

Another mysterious legend may provide clues to a tsunami. It is the story of the lost city of Atlantis.

Ancient myths describe a beautiful city under the ocean. People today still try to find it. And many scientists believe it exists.

The theory is that a volcano on the Greek island of Santorini erupted in 1628 B.C. The central dome exploded, throwing volcanic material into the air. It cooled into ash and rocks. Then it slid into the water.

The landslide caused a giant tsunami. The wave drowned

the island and its cities. Only bits of the volcano's crater rim are left. Where there used to be one big island, now only a ring of small **islets** remain. Some think this island's destruction began the legend of Atlantis.

At about the same time, the Minoan civilization on the island of Crete vanished. Crete is close enough to Santorini to be hit by the tsunami. Some think the same tsunami may have destroyed the Minoan civilization.

Why do scientists think a tsunami roared ashore at Santorini? What are the clues?

Scientists have found records on the rocks. Flood marks lie 300 feet above sea level along the Greek coastline. A tsunami could have made those flood marks.

Also, ancient Egyptians living at that time recorded events. They wrote that an island had exploded to the west. The sky had turned black. The sun disappeared. Streams and rivers filled with ashes. **Pumice** clogged the sea. Great waves rose and fell on the land.

Scientists think the recorded events sound like the Santorini explosion. And the waves sound like a tsunami.

Digging on the islands has turned up many clues. People have found signs of an advanced civilization under layers of pumice. Many scientists think this proves that the Lost City of Atlantis existed.

Today, visitors to Santorini find only the ring of islets. The ancient volcano lies in the middle under the ocean. The center of the volcano is 32 square miles. Imagine the huge explosion! One scientist compared it to a 1,000-megaton bomb exploding.

## What happened in Lisbon, Portugal, in 1755?

Another famous tsunami rolled into Lisbon, Portugal, in 1755. On November 1, a massive earthquake shook the port city. Crews on boats in the harbor stared. Buildings turned to rubble. People began screaming. A huge dust cloud rose, blotting out the sun.

Panicked victims trampled others. Terror reigned. Fires broke out everywhere. The bloody, shaken survivors rushed to the harbor. They hoped to escape by sea. The earth heaved again. The docks collapsed. Many more drowned.

Blazing fires raged through the city. A hot wind fanned the flames. To get to the water was the people's only hope. To escape in a boat would save them. Or so they thought.

Thousands of people crowded at the waterfront. They hoped to escape.

Then water began to drain from the harbor. The whole bottom of the harbor appeared. Boats heeled over. Hundreds of wriggling fish lay exposed. Rocks never seen before glistened wetly in the sun. The onlookers could see the ocean bottom a mile out.

Everyone stared, fascinated. They didn't realize that it was a deadly sight.

No one knew what it meant. No one knew about tsunamis.

No one knew what danger lurked. The earthquake had been their sign. But they had no idea.

Only minutes passed before the first wave came. Some said it was 50 feet high. The people of Lisbon saw the wave rushing toward them. Death would be certain.

Thousands panicked. They stampeded back into the flames. But the blazing heat drove them back to the waterfront.

People stood at the waterfront, helpless. The tsunami raced toward them. Its waters rose higher. Taller than a five-story building, it roared into the harbor. It threw huge ships against shattered buildings. It drowned and crushed thousands.

The tsunami surged more than half a mile inland. The water covered large parts of the city. Many who survived the first wave got caught in the backwash. They drowned. The death toll was enormous.

Two more waves struck. Hardly anything was left to ravage. Fifty-three palace buildings were gone. Thirty-two churches were destroyed. Only 15 percent of the city's 20,000 houses remained standing. And many of them were damaged even though they were on higher ground. The death toll was estimated at 20,000 to 30,000.

## What happened in Krakatoa in 1883?

Another famous tsunami occurred in 1883. A movie was even made about this. The island of Krakatoa exploded. Krakatoa is in Indonesia, within the Ring of Fire.

In the summer of 1883, undersea plates began rocking. Over a three-month period, Krakatoa's huge volcano erupted.

Volcanic ash drifted everywhere. A ship's captain reported sailing through huge masses of floating pumice for 500 miles. He was 1,000 miles away from Krakatoa.

Then small tsunamis began appearing. They were five to six

feet in height. On August 27, the first great explosion happened. A 400-foot peak disappeared.

In less than half an hour, a second explosion rocked the island. Millions of tons of seawater poured into the white-hot depths of the volcano. Steam pressure began to build.

These two explosions triggered tsunamis. Thousands drowned on nearby islands.

Then the volcano exploded for the third time. People thousands of miles away in Australia heard it.

More dust and ash exploded into the sky. It began drifting around the earth. The sky was so filled with volcanic material that people had to light their lamps in daytime.

The third volcanic eruption started a third tsunami. The central dome of the volcano blew. All the land above sea level was thrown into the air. The explosion threw much of the land below sea level into the air too. The land crashed down into the water again. These events triggered a huge tsunami.

This tsunami began rolling to the Bay of Merak. The bay channeled the tsunami and forced it even higher. Scientists guess the wave was over 100 feet high. A 33-foot wave struck Anjer Lor. A 70-foot wave demolished Tyringin. An 80-foot wave crushed Katimburg, 19 miles away from the Bay of Merak.

Krakatoa's destruction was enormous. Five thousand boats and ships were lost. Even boats in harbors hundreds of miles away sunk.

The waves circled the globe for days. Besides the bigger towns, almost 300 smaller towns flooded. Over 1,000 native villages flooded. The total death toll stood at 36,000.

## What happened in Japan in 1896?

Japan's worst tsunami was in 1896. An undersea quake rocked the coast. Sanriku, on the coast of Japan's main island, Honshu, felt it.

But fishermen fishing in the ocean didn't notice the tsunami beneath their boats. They had no idea what was happening in their villages.

The fishermen returned to find their homes destroyed. Their families were gone. An 80-foot wave had destroyed their villages. Over 27,000 people died.

## What happened in Hilo, Hawaii, and Alaska in 1946?

In 1946, a famous tsunami attacked Hilo, Hawaii, and Alaska. It began as an earthquake in Alaska.

On April 1, 1946, a 7.3 earthquake rocked the ocean bottom. It was near the Aleutian Islands off the coast of Alaska.

Within 45 minutes, death struck in Alaska. A 100-foot-high wave smashed into Scotch Cap lighthouse. This lighthouse stood on land 32 feet above sea level. It was built of concrete and steel and was supposed to handle anything. The tsunami pulverized it.

Five Coast Guard men died. Their bodies disappeared.

April 1, 1946. Tsunami strikes the beachfront area east of Hilo, Hawaii.    NOAA

April 1, 1946. Waves wash up Waiauenue Avenue in Hilo, Hawaii.

Rescuers found a piece of human intestine on a hill behind the site. A bronze tablet marks the spot today.

But the tsunami wasn't finished. It began racing across the Pacific. Five hours later, at 7 a.m., it arrived in Hilo, Hawaii. A few smaller tsunamis hit before the big one.

Wallace Young owned a clothing shop in Hilo. Someone phoned him to say a small wave flooded the street in front of his shop.

Mr. Young hurried to his shop. He checked for damage. Others did the same. Then a policeman ran down the street. "Here it comes!" he yelled.

Mr. Young looked out a second-story window. He couldn't see past the two-story buildings across the street. Then he saw something. The buildings across the street were moving. They were moving right toward him. A couple was hanging onto the ledge of a window.

The wave's force bent parking meters in half. Gene Wilhelm, reporter for the *Hilo Tribune-Herald,* wrote about the power of the large wave.

"Whole buildings were driven directly through the ones behind, the way you'd close a telescope. Boulders weighing nearly a ton were picked up and set down hundreds of yards away. And through it all, there was the sickening thunder of buildings disintegrating, with the screams of people trapped inside."

The tsunami ripped up railroad tracks. It buried coastal highways. It washed away beaches and killed 159 people. Afterwards, the ocean was filled with floating houses.

Hilo, with its funnel-shaped harbor, is a natural target for tsunamis. The narrowing channel forces the tsunami waves even higher.

Engineers have tried to think of ways to protect it. They talk about building a gigantic breakwater. They talk about raising waterfront buildings on rods of steel embedded in rock. But the costs are too high.

Today, in Hilo, a park lies next to the waterfront. But the town was rebuilt much farther inland. People in Hilo know the power of the death wave.

It was after this disaster that the Pacific Tsunami Warning Center began. No one wanted destruction like that again.

## What happened in Alaska in 1958?

In 1958, a landslide caused a tsunami in Alaska. It is the largest landslide tsunami ever recorded.

On July 9, Bill and Vivian Swanson anchored their boat just outside Lituya Bay. That morning, they felt the boat rock. It was an earthquake.

They looked across the bay at the mountains. Suddenly, about 40 million cubic yards of rock plunged down from the mountains into the bay. The splash roared up the hillside on the other side of the bay. It splashed over 1,700 feet high. (That's 500 feet higher than the Empire State Building!)

Trees on the mountains were uprooted. The wave stripped the soil down to bedrock.

Then the giant wave began heading for the Swansons. They could only watch in shock and terror. The wave roared across the bay at more than 100 mph. It took four minutes to reach them.

Their boat shot up to the wave's crest. It began to surf backward. Bill Swanson looked down to see the tops of trees 80 feet below him.

The wave swept them into deep water. The bow pointed up. Luckily, air was trapped inside the boat's hull.

The Swansons were able to get into a smaller boat they had. Rescuers found them two hours later.

As they searched the bay, scientists discovered flood marks from ancient tsunamis. At least four ancient tsunamis had roared through the bay.

The Tlingit Indians, who once lived by the bay, had a legend they told over and over. They talked about a sea monster who lived beneath the waves. They thought he sent huge waves to keep people away from his land.

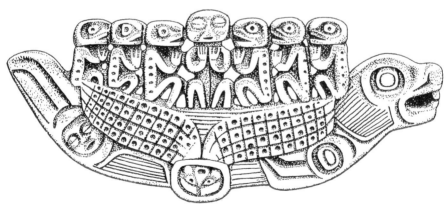

Tlingit Spirit Canoe

March 27, 1964. Kodiak, Alaska. Tsunami caused 21 deaths and 30 million dollars in damage in and near the city of Kodiak.

## What happened in Alaska in 1964?

In 1964, a huge earthquake shook Alaska. The earthquake was an astounding 8.5. Its power was 12,000 times greater than the atom bomb dropped on Hiroshima.

Karl Armstrong, the editor of *The Kodiak Mirror,* wrote that when he tried walking, it was like "marching across a field of jello."

Tons of earth fell into the sea along the Alaskan coast. Waves over 90 feet high washed over coastal towns.

This tsunami was a smooth, rapid rising of the sea. It never broke in a **crest**. Silently, it brought death. The second wave roared in at 30 feet high, bringing destruction with it.

The tsunami also hit Crescent City, California. Two waves washed into two streets. Then they pulled back out to sea.

Several people thought it was safe to go back. They wanted to see what had happened. But a third, huge wave raged in. It carried them out to sea. One hundred and twenty-two people died in the 1964 Alaskan tsunamis.

## What happened in Nicaragua in 1992?

In 1992, two huge tsunamis brought death to the Pacific Rim. In September, a "slow-slip" earthquake hit the Pacific coast of Nicaragua.

In chapter 3, you read about research recently done on these "mild" earthquakes. At that time, no one thought anything bad would happen. But they were wrong.

Two Americans, Chris Terry and his friend Scott Willson, were on their fishing boat. They were anchored in San Juan del Sur's harbor. They had felt the mild earthquake earlier. But they thought nothing of it.

They were both down below deck when "we heard a slam," Terry says. The bottom of their boat had just scraped the bottom of the harbor.

The harbor, which was usually more than 20 feet deep, had just drained. It was as if someone had pulled a giant plug.

Then a powerful wave lifted up their boat. "Suddenly, the boat whipped around very, very fast," says Terry. "It was dark. We had no idea what had happened."

The boat began dropping again. It dropped into the trough of a giant wave. Willson got up on deck. They were being

Tsunami damage at El Tranisto, Nicaragua. Sixteen people were killed, 151 injured, and more than 200 houses destroyed. NOAA

carried on a giant wave toward the shore. The back of a huge wall of water stared him in the face. "And then the swell hit, and the lights went out, and we could hear people screaming," Terry says.

After the tsunami receded, death and destruction were left behind. The waves swept away restaurants and bars lining the beach. Homes, cars, and people had been carried hundreds of yards inland. "When the wave came back out, it was like being in a blender," says Terry. Collapsed homes floated in the water around their boat.

These waves were destructive. But they were only five to six feet high at San Juan del Sur.

Elsewhere in Nicaragua, the waves were at least 30 feet high. They killed about 170 people, mostly children who were sleeping. More than 13,000 Nicaraguans were left homeless.

## What happened in Indonesia in 1992?

Indonesia was also hit by a tsunami in 1992. An earthquake off Flores Island triggered deadly waves. Whole villages washed out to sea. Over 1,900 people died. Forty thousand people were left homeless. Over 21,000 homes were destroyed. Several hundred churches, mosques, and other buildings became rubble.

Most of the damage was from the tsunami. Some witnesses said they saw 80-foot waves. The waves left dead tuna in village streets and washed bodies of victims out to sea.

December 1992. Leworahang, Indonesia. A paved road comes to an abrupt halt at the edge of the cliff. NOAA

July 12, 1993. Aonae, Okushiri Island, Japan. A fishing boat is beached near a damaged fire truck.

NOAA

## What happened in Japan in 1993?

In July of 1993, an earthquake in the Sea of Japan sent a tsunami to Japan. It was one of the largest ever to hit that country. Waves reached to over 97 feet above sea level. Over 120 people were killed.

## What about the future?

Tsunamis rage over the ocean. They ravage the land. They destroy everything in their paths. As long as people live near oceans, tsunamis will devastate their lives.

Luckily, tsunamis don't happen very often. About five tsunamis are reported every year. Most are small. Destructive tsunamis seem to hit about once every ten years.

But the earth will keep shifting. Earthquakes, volcanoes, and landslides are a fact of life. So tsunamis will continue to happen.

People must be aware if they hope to survive. They shouldn't take risks. They shouldn't be foolish. They must be ready. Only then can people hope to escape the death wave— the tsunami.

March 28, 1964. This damage was caused by the tsunami near Seward, Alaska.     NOAA

# GLOSSARY

**civil defense**  a system of protective measures and relief activities in the event of an attack or natural disaster

**core**  very hot rocks deep in the center of the earth

**crest**  the highest portion of the wave just before it breaks. Its height is measured in comparison with the ocean's level around it.

**crust**  what scientists call the topmost layer of the planet earth. It is made of metal, soil, and rock. In some places, it is 20 miles thick; in others, 40 miles thick. It "floats" on the mantle of molten metal below it.

**earthquakes**  a geologic event where plates in the earth's crust slide past each other or slide on top of each other

**geologists**  scientists who study the earth

**geology**  the study of the earth

**gravity**  the force that pulls everything toward the center of the earth

**islets**  small islands

**landslide**  the sliding of a mass of rocks or earth down a slope

**mantle**  a layer of what may be molten metal under the earth's crust. In some places, it is 1,800 miles thick, scientists think.

| | |
|---|---|
| **National Earthquake Information Center** | (N.E.I.C.) a center where scientists study and track earthquakes |
| **oceanographer** | a scientist who studies the ocean and its movement |
| **Operational Tsunami Warning System** | a system set up by the Intergovernmental Oceanographic Commission that studies and tracks tsunamis around the world and issues warnings to affected countries. The Alaska Tsunami Warning Center and the Pacific Tsunami Warning Center are its main stations. |
| **Pacific Rim** | the ring of countries surrounding the Pacific Ocean |
| **plates** | the term scientists give to the layers of soil, rock, and metals covering the earth. They are in huge pieces and can move or drift. |
| **pumice** | porous, lightweight volcanic rock |
| **Richter scale** | the measuring device used by scientists to measure how strong earthquakes are |
| **Ring of Fire** | the name given to the countries circling the Pacific Ocean. It follows the western edges of North and South America, loops under the ocean, and follows the coasts of Australia and Japan up to Russia. |
| **seismic** | caused by an earthquake or earth vibration |

| | |
|---|---|
| **slow motion slip** | earthquake movement caused by long-period waves. This movement is not as noticeable as short-period waves that jolt and jerk the earth. |
| **sonar experiments** | scientific experiments done by bouncing sound off an object and measuring how long the sound takes to return to its source |
| **storm surge** | waves whipped up by high winds in a storm |
| **subduction zones** | where one plate overrides or subducts another. The lower plate is pushed down into the mantle. |
| **submarine canyons** | canyons underwater in the ocean's floor |
| **tidal waves** | waves caused by tides. Gravity causes tides. Tides are regular and have rhythm. |
| **tsunami** | a huge movement of water caused by a geologic event—an earthquake, a landslide, or a volcanic eruption |
| **tsunami earthquake** | an earthquake that causes a tsunami |
| **volcano** | a vent in the earth's crust through which molten lava and gases are poured, or a mountain formed by the lava |

# Index